ELEGANT EX

ELEGANT EXPLORATIONS
THE DESIGNS OF PHILLIP JACOBSON

Grant Hildebrand

UNIVERSITY OF WASHINGTON PRESS

SEATTLE AND LONDON

NORDIC HERITAGE MUSEUM

SEATTLE

Publication of *Elegant Explorations*
was made possible in part by generous
support from the Department of Architecture,
College of Architecture and Urban Planning,
University of Washington.

University of Washington Press
P.O. Box 50096, Seattle, WA 98145 U.S.A.
www.uwashington.edu/uwpress

Library of Congress Cataloging-in-Publication Data
Hildebrand, Grant, 1934–
Elegant explorations : the applied designs of Phillip Jacobson /
Grant Hildebrand
p. cm.
Includes index.
ISBN-13: 978-0-295-98719-4 (pbk. : alk. paper)
ISBN-10: 0-295-98719-7 (pbk. : alk. paper)
1. Jacobson, Phillip L. 2. Architect—designed decorative arts—
United States. I. Title.
NK1412.J24H55 2007
745.2092—dc22   2007009529

The paper used in this publication meets the minimum
requirements of American National Standard for Information
Sciences–Permanance of Paper for Printed Library Materials,
ANSI Z39.48-1984

Frontispiece:
Wood Log, San Juan Islands. Washington, 1970

FOR EFFIE

CONTENTS

Eroded Sandstone
Canyon de Chielly, Arizona, 1986

FOREWORD

Our technological age is an era of excessive professional specialization. Knowledge and skills have become increasingly focused and limited in their scope. At the same time, however, it has been observed that new innovations often emerge at the boundaries between the various disciplines. Similarly, insights tend to occur in unfocused mental states when the conventions of the accepted disciplinary paradigm momentarily lose their validity and grip.

Today, the discipline of architecture is usually understood as a strict professional practice aimed at merely producing utilitarian buildings. Historically, however, the art of building is rooted in a wide base of human knowledge ranging from cosmological and philosophical interests to engineering skills, crafts, and the arts. Significantly, one of the greatest architects of all time, Philippo Brunelleschi, was initially a goldsmith and clockmaker. The modern era still has produced great generalists; a single person could work as an urbanist, architect, furniture designer, sculptor, painter, and writer, as did Le Corbusier. Or an architect could design everything from a city hall and factory through interiors and furniture to kitchenware, cutlery, and bathroom fittings, as did Arne Jacobsen. Regardless of today's tendency toward specialization, architecture continues to be a discipline that bridges C. B. Snow's two categories of human knowledge: the technological and the humanistic, the scientific and the artistic. This is the true fascination of our profession.

As such, architecture does not merely give rise to material buildings; our constructions and artifacts structure our per–ceptions and life experiences by providing distinct horizons of scale, measure, and meaning. We live in a world that is pre-structured by human constructions. When architecture is understood in its existential meaning as a mediator of human experience, architectural thinking expands naturally beyond specific tasks, functions, or scales.

Architecture is essentially the art form of relatedness. A building is always related to its landscape or urban context and specific patterns of movement or use. There are also internal relationships within the architectural work itself between large and small, outside and inside, space and form, entity and detail. In architectural design, forms and assemblies are usually generated by principles and systems of variation and articulation. Forms in architecture tend to be constructed and assembled, rather than sculpted.

Thus there is a distinct difference between the product designs of architects and those of designers. The architect's inherent interest in the orchestration of ensembles of spaces, structures, and details naturally leads to an application of aesthetic principles or aesthetic genetic codes, as it were, through various scales. Design products of architects are frequently born in conjunction with particular architectural projects, and consequently they echo the aesthetic logic of their "parent object."

There are two different approaches to design. The first focuses on the object itself, the second is concerned with the dialogue of the object and its physical as well as cultural context. One designer aims at conceiving the object as a closed and finite aesthetic entity, whereas the other sees the designed product as an ingredient in and a dialogue with the settings of life. One aspires to draw attention to the object and focuses primarily on aesthetic expression, the other considers these matters within the larger context of the realities and poetics of life.

There can be no doubt that Phil Jacobson's values are oriented toward the latter preferences. As an architect educated in modernist aesthetics and ethics and deeply influenced by the Nordic and Japanese design traditions, he is also bound to value simplicity, reduction, and restraint. He is inclined to think of visual relationships rather than singular objects, of applicable aesthetic systems instead of idiosyncratic individual objects.

His object designs are a form of miniaturized architecture. He extends the realm of architectural thought from the scale of urban strategies and complex building projects through interior spaces, furniture, light fixtures, and serving pieces, to the micro scale of jewelry. His design process is based on modularity and careful articulation of proportions, as well as clarity of structure, assembly, and detail. The logic of his process makes his work resonate with architectural images; a lamp turns into a tiny building, a sconce into a unit of modular construction, a ring into a miniature castle. His images are not fixed motifs; the associations keep shifting from one image to the next. A pendant appears as a stage, a display case, or a window frame—it may also evoke associations with architectural stairways or stepped pyramids. His tea and coffee set suggests a fortified city, and his wife's belt buckle turns into a delicate façade of sterling silver. Every great artistic work—poem, painting, film, building, or piece of jewelry— creates its own microcosm, an autonomous world. The city and the house, the piece of furniture and the bracelet are all metaphors of an idealized world.

Jacobson's passion for photography has trained his eye to the infinite richness of geometric patterns as he has found them in various cultures around the world. It has also trained his eye for proportion. He is particularly fascinated by the Golden Section, or the Divine Proportion, the mythic and controversial system of proportion that emerged in the theories of the Italian Renaissance. So too, one set of his pendants is based on the primary geometric forms of triangle, square, and circle. These are the archetypal elements of ornamental pattern—but they are also symbols of magic and elements of the famous calligraphic painting *Universe* by the seventeenth-century Japanese Zen artist Sengai, as well as the ingredients of modernist abstract compositions since the Bauhaus. Every profound artistic form contains a multitude of images.

An element of play and carefree experimentation is an essential ingredient of all profound creative search. Without the liberating attitude of play and the excitement of unexpected discovery, architectural practice threatens to degenerate into a joyless routine. As actual architectural tasks are always severely restricted by functional requirements, structural and technical demands, and social responsibilities, the architect needs to shift his mode and scale of exploration occasionally to the field of artistic experimentation, or formal play. Experimentation produces aesthetic and intellectual concepts that can later be applied in actual architectural design work. Most importantly, it reminds the designer of the poetic dimension of architecture.

As an accomplished and respected teacher, Phil Jacobson sets a valuable example of an architectural exploration into the miniaturized world of objects. Jacobson's shifting of scales and foci has surely invigorated his own architectural work; it may well inspire others to step beyond the boundaries of their professional routines.

JUHANNI PALLASMAA
Helsinki, October 2006

ELEGANT EXPLORATIONS

El Morro Fortress,
San Juan, Puerto Rico, 1989

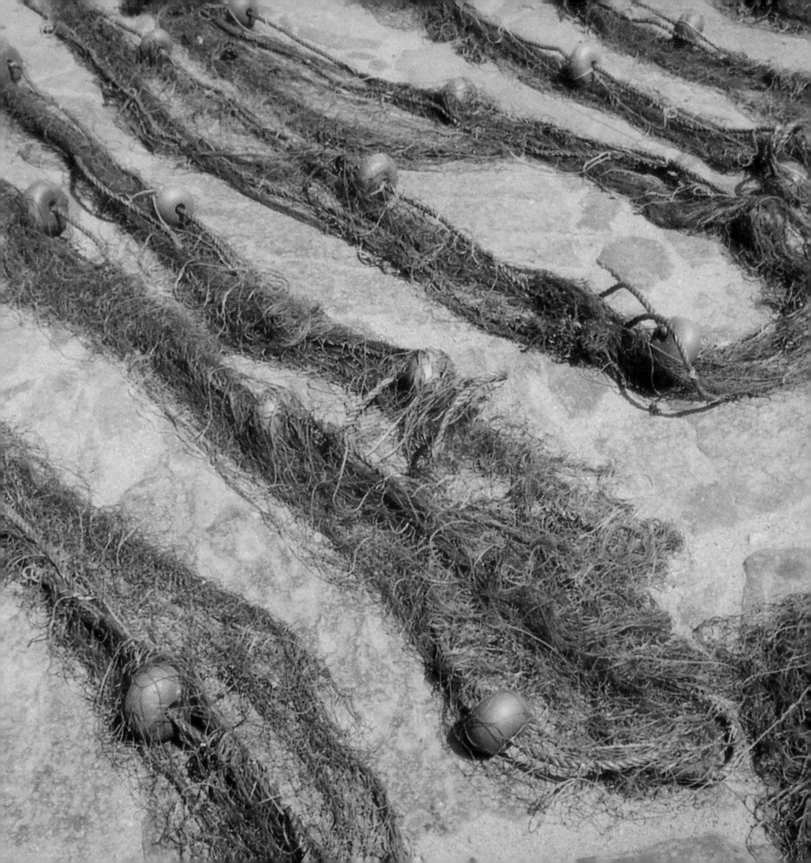

BACKGROUND

Allen Jacobson, Phillip Jacobson's father, was a master constructor of prototype aircraft for stress testing and test flying. After an early career as superintendent of experimental aircraft for Douglas in Los Angeles, in 1934 he assumed a similar position at Lockheed, directing the building of its experimental aircraft, including the P-38 fighter and the Lodestar passenger plane. In 1941 he came to the Boeing Company in Seattle, where he was responsible for prototypes of the B-29, B-47, and B-52 bombers and the Stratocruiser passenger airliner. Those positions testify to Allen Jacobson's skills and standards; Phillip remembers that "my father was a master craftsman. . . . He accepted nothing but perfection and expected the same from those with whom he was involved. He seemed to me the exemplar of an innate emotional commitment to excellence, and he imbued in me a sense of the joy inherent in the act of making. His influence on me was both profound and lasting."

Washington State Convention and Trade Center, Seattle, 1988 "A significant work of urban architecture that embraces the city instead of defying it."—Paul Goldberger, *New York Times*

After graduation from high school in Seattle in 1946, Jacobson spent a year with the Army's occupation forces in Japan, where he discovered his interest in architecture, and "experienced the wonders of indigenous Japanese folk art and crafts." When he returned to Seattle in 1948, he had decided to study architecture. Though the University of Washington was in his home town, he went to less-crowded Washington State University in Pullman. The curriculum there was heavily oriented toward the Bauhaus methodology, as was typical of American schools of architecture at the time. Jacobson, therefore, would have been

Mahlon Sweet Airport, Eugene, Oregon, 1980
Forms and spaces that create a sense of flight rendered in materials of local significance.

Fishing Nets
Marseilles, France, 1952

deeply immersed in the philosophies of European modernism, especially as represented by Walter Gropius and Mies van der Rohe. Yet his subsequent actions and observations indicate that he was also aware of a broader range of contemporaneous work and ideas, notably that of several less well-known Scandinavian designers.

In the fourth of Jacobson's five years at Washington State, he met Effie Galbraith, a student in sociology who had studied in Oslo the previous summer. She encouraged him to go to Europe; he followed her advice. Immediately after graduation (with honors) in 1952, he spent four months traveling from northern Norway to southern Italy, then nine months, on a Fulbright grant, at the University of Liverpool, for graduate study in urban design and planning. He then explored Europe for another three months. He remembers the time as "the finest year I ever had in terms of intellectual growth."

During those months of travel, historic architecture took on a whole new meaning for me in its artistic quality, its social context, its technical attainment, and its humanism, fanning in me a passion that has steadily increased in ensuing years. This was also the opportunity to experience at first hand the work of some of those modern architects and engineers I had studied from afar: Le Corbusier, Robert Maillart, Peter Behrens, Pierre Chareau, Alvar Aalto, Gunnar Asplund, Sigurd Lewerentz, and Arne Jacobsen. I was impressed that most of these architects were involved in an extraordinary range of design. This was my first experience with complete modern architectural environments in which the building architect had also designed the lighting, furniture, hardware, and other items specifically for a particular building. I especially admired the work of the Nordic designers, their dedication to comprehensive and humanistic design of astonishing

GHI Wing, University of Washington, Seattle, 1994

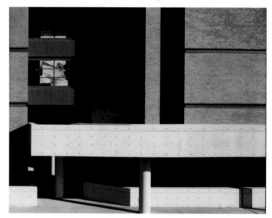

Biological Science Building
University of Washington, Seattle, 1980

4

quality. That admiration would remain with me, as a continuing influence on all of my work.

He returned to Seattle in late 1953. Effie, having encouraged him to travel, had patiently stayed at home throughout his wanderings. They were married on his return.

After a short time with another Seattle office, Jacobson accepted work with the firm founded and headed by Steve Richardson that would later become The Richardson Associates, then TRA. But Jacobson also wanted experience with urban projects of larger scale than was then available in Seattle. He and Effie returned to his native California, where for two years he worked for John Carl Warnecke in San Francisco. Richardson then persuaded Jacobson to return to TRA, where he became a project designer, associate, then partner and, for more than twenty years, design director, until his retirement from the firm in 1992. During his tenure at TRA it would become a large multidisciplinary design firm, widely recognized for its deft and effective design of large, complex architectural and urban design projects, both domestic and international. His firm received more than 130 national, regional, state, and local design awards, and its work was published in over 80 different periodicals and books, foreign and domestic. Characteristic examples of his architectural work are the Biological Sciences Building for the University of Washington in 1980, the Washington State Convention Center in Seattle in 1988, the Albuquerque Airport in 1989, the King County Aquatics Center in 1990, and the Jacobsons' own house in the Laurelhurst district of Seattle in 1972.

In recognition of his accomplishments in architecture Jacobson was elected Fellow of the American Institute of Architects in 1973, and in 1994 he received the AIA Seattle

King County Aquatics Center, Federal Way, Washington, 1990. One of the fastest swimming facilities in the world. Site of the International Goodwill Games.

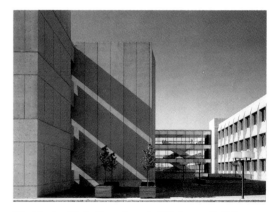

Washington State Department of Highways Building Olympia, 1970. A pioneering effort in the use of multistory, precast elements to create a formal relationship to an existing classical governmental complex.

Medal, the highest honor the Seattle AIA confers, for distinguished lifetime achievement.

From 1962 to 2000, Jacobson also taught at the University of Washington, in the departments of Architecture and of Urban Design and Planning, where he offered graduate and undergraduate design studios to more than a thousand students. That he guided over 150 graduate theses, far more than any contemporaneous faculty member, testifies to the students' consistent admiration for his teaching skills. At various times in that period he held visiting appointments at the Tokyo Institute of Technology, The Royal Academy of Fine Arts in Copenhagen, The University of Sydney, and The Royal Institute of Technology in Stockholm. He has continued to travel often and extensively to Europe, Scandinavia, Mexico, Japan, and the Middle East and North Africa, where he found an appreciation for Islamic arts.

Jacobson's early admiration for the extraordinary quality of Nordic design was confirmed and enlarged during a sabbatical year in Helsinki in 1968, in which he was awarded a degree of Master of Architecture from the Finnish Institute of Technology. For his continuing work toward strengthening design relationships between Finland and America and in recognition of his architectural achievements, he holds the Silver Award from the Finnish Society of Architects, and the Government of Finland has created him Knight First Class of the Order of the White Rose.

In parallel with this known and recognized career, however, Jacobson has continuously engaged another realm of design. The comprehensively designed environments he encountered in his early travels through Europe and Scandinavia have led him to an interest in small-scale design that has encompassed almost the entirety of his

Seattle City Light, South Service Center, Seattle, 1968.
An early experiment in combining precast, prestressed and post tensioned concrete elements, all combined in one structure.

Jacobson Residence, Seattle, 1972
"It was like living in an art museum but not having to pay admission."—Erik Jacobson

career and continues to this day. He has designed furniture and lighting fixtures for some of his firm's wide range of projects and has seen those designs through production. He has designed furniture and lighting fixtures, too, for his own home and family. And almost from the outset, he has designed jewelry and home accessories for friends, family, and above all for Effie. That work, like his architecture, has been founded in Jacobson's continuing intention to create designs of enduring quality, whose delight derives from timeless structural and formal concepts.

That realm of design has been for Jacobson an exciting exploration. It has also been an exciting challenge because, whatever the parallels, the creative process entailed is shaped by considerations different from those that govern architecture, is often largely driven by subjective personal insights, and is sometimes exclusively obedient only to the designer's own imagination. That other realm of Jacobson's life is the subject of this book.

Albuquerque International Airport
New Mexico, 1989. The essence of the region's indigenous architecture embodied in contemporary form.

II FURNITURE

Jacobson's earliest years in architecture coincided with an extraordinarily creative period of furniture design, much of it by architects. It was the era of Eero Saarinen's Womb chair and his Pedestal series; Harry Bertoia's many pressed steel mesh chairs; Isamu Noguchi's famous free-form glass-topped coffee table; and the many designs for tables, storage units, and above all chairs, by Ray and Charles Eames. That was also the era of greatest creativity of the Nordic designers, foremost among whom were, in Denmark, Hans Wegner, Finn Juhl, and Borge Mogensen, and in Sweden, Bruno Mathsson, Carl Malmsten, and Yngve Ekström. Although designed much earlier, Mies van der Rohe's many furniture pieces, and an equal number by Marcel Breuer, were being produced in this country, and Alvar Aalto's lovely and timeless designs, dating mostly from earlier decades, were also available.

The work of these designers seems diverse; Wegner's famous oak dining armchair would seem to have nothing in common with Mies's Barcelona chair. But all reveal the assumption that the materials of which the piece was made should be evident; it is hard to think of a single example in which paint is used as a finish. All exhibit the means of construction and structure. All renounce historic precedent. All avoid applied ornamentation. All, at a conceptual level, accomplish their purposes with a minimum of means—no

or even efficient production; both Mies's chair and Wegner's are costly indeed. Yet it must be said too that many pieces were seriously designed to be produced inexpensively by new technologies, in new materials—the Eames fiberglass chair shells, or Saarinen's Pedestal series, come to mind. And among the Scandinavian designers there was often, and especially in the design of case goods, a dedication to designing for easy and economical knock-down shipping.

A MODULAR SEATING SYSTEM 1974

Jacobson's first serious work in this realm dates from 1974, with the design of a modular seating system. It is clearly consistent in every way with the design precepts of the era. It acknowledges no historic precedent; it includes no ornamental flourishes. All units are dimensionally coordinated to allow an all-but-infinite variety of arrangements. In a fully upholstered piece, of course, the question of evident means of construction and structure does not arise. The only materials that meet the eye, however, are luxurious. All exposed surfaces are of Brazilian black luggage-stitched leather; the connectors, low on the backs and sides, that secure those panels to the chassis are of satin-finished stainless steel.

A DINING TABLE, SIDEBOARD, AND
TWO COFFEE TABLES 1978

These pieces were designed as a coordinated ensemble. Serving surfaces for all four pieces are of black plastic laminate over high-density particle board, the edge is solid beech. The dining table cantilevers from a chromed steel-tube and steel-plate trestle to allow complete freedom of seating around the perimeter; the steel trestle telescopes to allow insertion of a central leaf. The other three pieces are supported by rectilinear elements of chrome-plated steel bar stock. The design was driven by the desire for an element that would be functionally and aesthetically appropriate to both a wall-mounted bracket and a table support.

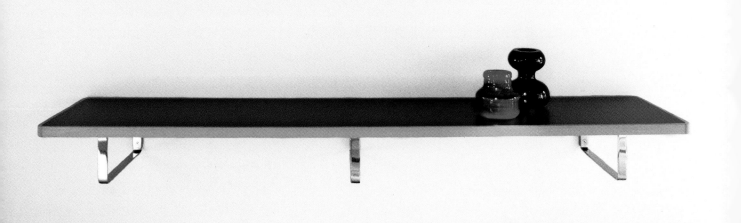
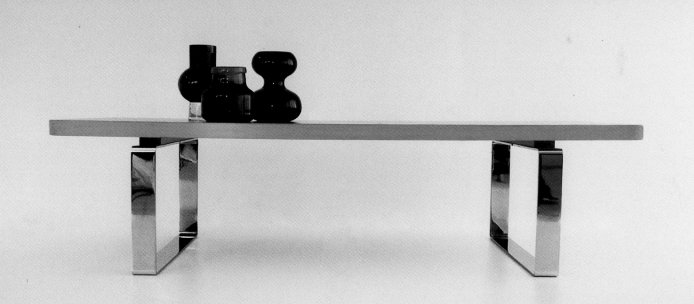

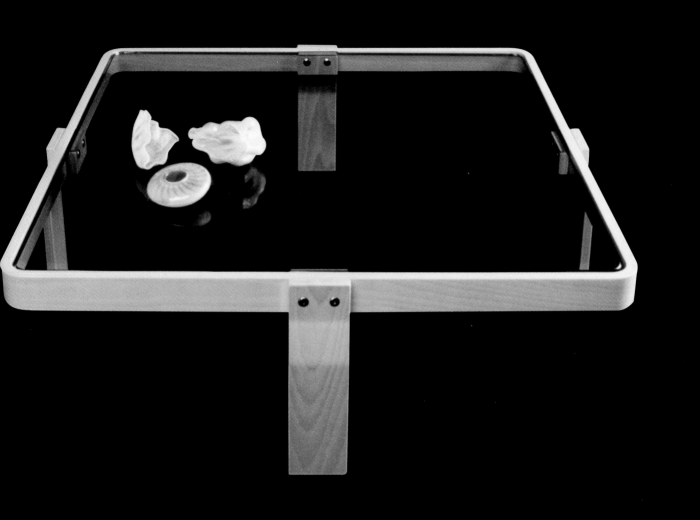

THE LAMINATED WOOD SERIES:
A COFFEE TABLE 1990

The coffee table's frame is of beech-veneered plywood; the shapes are formed and locked in at the time of lamination. As such it, like the two chairs and the counter stool to follow, is a descendant of many twentieth-century designs for laminated wood—plywood—including Aalto's and Breuer's lounge chairs of the 1930s, the Eames's dining chairs of the 1940s, and their classic leather-and-rosewood lounge chair of 1956. In most cases, in keeping with the precept that the materials of which the piece is made should be evident, the plywood edge was left exposed. This, in turn, demanded that most of these pieces were made of solid-core plywood to avoid the unsightly voids typical of the ordinary material.

Jacobson's designs in laminated wood are marked by certain distinctive features. Short zones of curvature occur within dominant areas of planar material. Curvatures are of small radius, and the radii bear close and evident relationships to one another; often a single radius is repeated throughout the piece. The forms of constituent parts, and the overall form of the assembled piece as a totality, tend toward the rectilinear; straight lines and right-angle relationships predominate. Parts are joined to one an-

other quite simply, and are seen to join, the broad side of one piece facing that of the other, with bolts or screws to secure the connection.

These characteristics serve economy of production and, in the case of the chairs, economy of shipping, since they are easily disassembled into parts capable of close packing. These characteristics also give Jacobson's designs a unique presence. The crisp geometries confer a dignity unusual in such seemingly modest pieces—perhaps not coincidentally they seem architectural—and the means of making, interesting in itself, is eloquently evident to the eye.

The coffee table was Jacobson's first exercise in this technology. The legs, whose outward reach is intended to yield a visually reassuring sense of stability, support a square horizontal frame that seems fragile for its role. But in fact its role is not what it seems. It locates and secures the legs and protects the edge of the half-inch plate glass top. But the edge of the glass lies slightly inboard of the wooden frame, and the glass is supported only on the ledger that occurs for that purpose at the juncture of legs and frame. The glass, therefore, cantilevers toward all corners, creating a subtle visual tension, while its very considerable weight, bearing directly at the tops of the legs, ensures the physical stability of the piece.

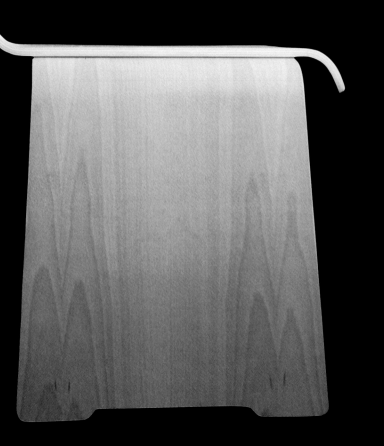

THE LAMINATED WOOD SERIES:
DINING CHAIR I 2004

The design of this chair is influenced by the Nordic development, since World War II, of furniture designed around the mandates of efficient packing for long-distance shipment. It is also inspired by recent furniture designs in plywood including, specifically, a footstool design for the Swedish home-furnishings dealer IKEA. The chair can be easily disassembled into its two constituent parts, the seat-and-back and the supporting side planes that, bent, continue under the seat. The geometry of each part allows close packing to conserve container volume. The chair is simply reassembled by means of four chrome-plated bronze bolts. The seat and back are designed with particular attention to ergonomically correct support, with a degree of resilience. Optional padded seat and back cushions in black leather can be attached with Velcro strips.

THE LAMINATED WOOD SERIES:
DINING CHAIR II 2004

This chair is a rather direct progeny of Jacobson's glass-topped coffee table, whose frame and leg configuration it borrows. A three-legged design has the advantage that it accommodates uneven floor surfaces without rocking. It also has the disadvantage that it is less stable; Jacobson's use of outward-splayed legs is intended to counter that tendency. The frame is again of beech-veneered plywood, with a black leather seat pad. There are five constituent pieces, connected by means of ten polished chrome-plated bronze fasteners; the pieces are again designed for easy disassembly and assembly and for close-packed shipment. In this case, however, each piece is separated from its neighbor by a thin spacer, creating a tidier joint and more clearly articulating the elements of the piece. The chair is intended for both domestic and commercial use.

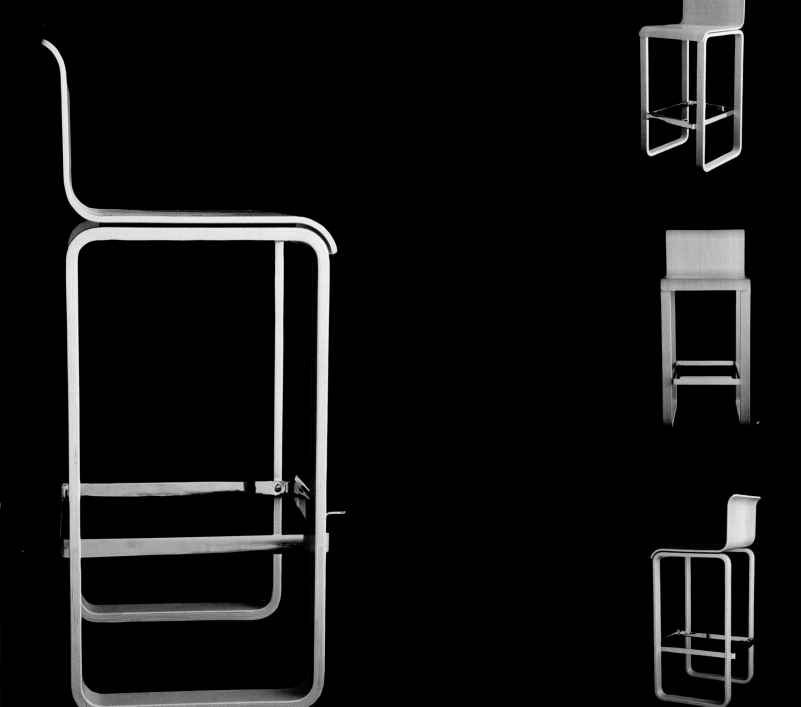

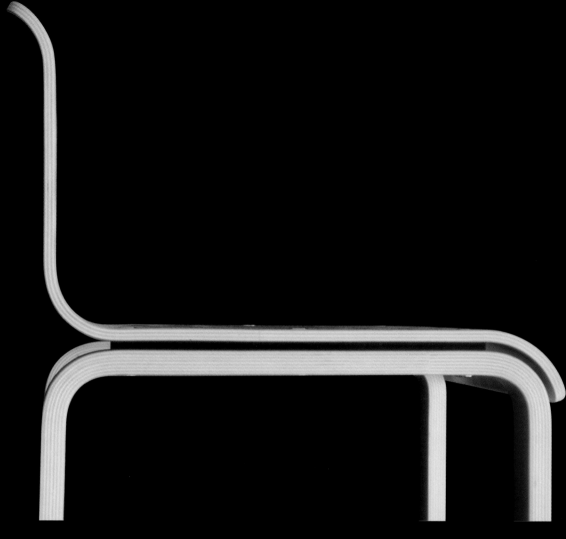

THE LAMIINATED WOOD SERIES:
A COUNTER STOOL 2005

A compatriot of the dining chairs, the counter stool is
also designed for efficient packaging for shipping; it is assembled with six bolts and four screws. The seat, back,
and legs are, again, of laminated beech; the footrest is
polished chrome-plated steel. The back is low but stiff,
for lumbar support. The dimensions of the floor contact
points are maximized for stability. The coordination of
the various radii is especially evident in this piece and is a
major source of its visual dignity.

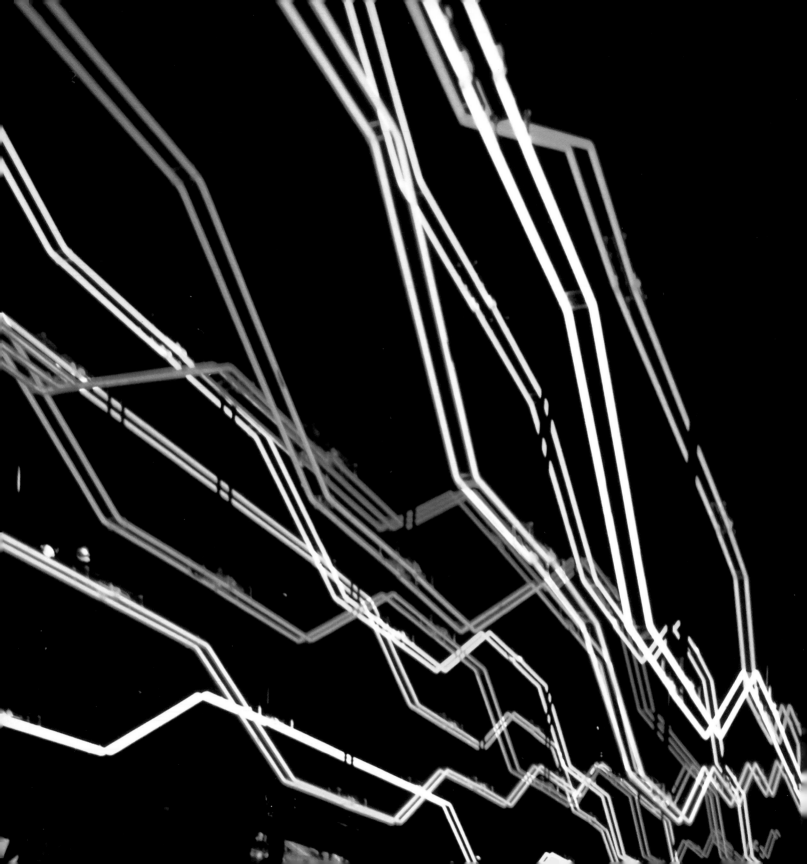

The design of an electrical lighting fixture is driven by several general considerations. The fixture must provide a level of illumination appropriate to the purposes it serves, but normally the actual light-emitting element should be hidden from view. If the enclosure is translucent, the brightness of its surface, for any given illumination, is inversely proportional to its area; thus size, for such a fixture, directly affects its performance. Fabrication and assembly techniques must be commensurate with the budget or, for a production unit, the reasonable market price. The fixture must be capable of straightforward installation; if a residential fixture it should, ideally, be capable of installation by an owner unfamiliar with professional installation practices. Finally, though this should perhaps be said at the beginning: since the human eye is naturally and unavoidably drawn to the brightly lit areas of any scene it surveys, light fixtures inevitably call attention to themselves. For that reason they are often intended to be, or become, key features of a setting. The grand chandeliers of the nineteenth and early twentieth centuries are familiar examples; so too is the neon ceiling of the concourse at Chicago's O'Hare Airport.

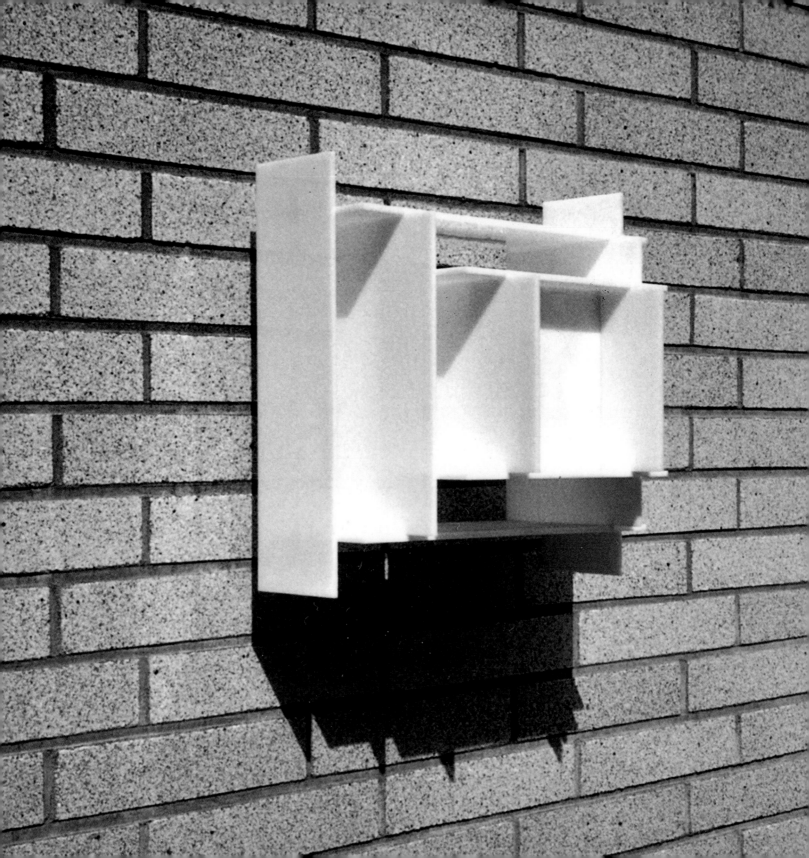

INSTITUTIONAL FIXTURES 1960–65

The period from 1945 to 1965 was a remarkably creative time in lighting fixture design. In those years the ubiquitous flex-armed Swedish Luxo lamp, in its many versions, became an icon of modern design. Poul Henningson was working out his still-produced classic series. George Nelson's range of lamps for Howard Miller, comprising a translucent plastic sheath over a wire armature, recalled the inexpensive and popular vernacular paper lamps from Japan. Isamu Noguchi even designed a charming example for table-top use. But most of the attention was given to high-end residential fixtures; of affordable well-designed fixtures for institutional use there was a dearth.

In the late 1950s, Jacobson chose to address the problem. Working with Tom Wimmer, President of the Seattle Lighting Fixture Company, in the company's manufacturing shop, he learned the basic characteristics of fabrication processes and materials and designed numerous fixtures that were mass produced for commercial and institutional use. The fixtures were made of enameled metal, welded acrylic, and, occasionally, wood, and were installed in schools, university buildings, and churches.

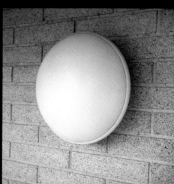

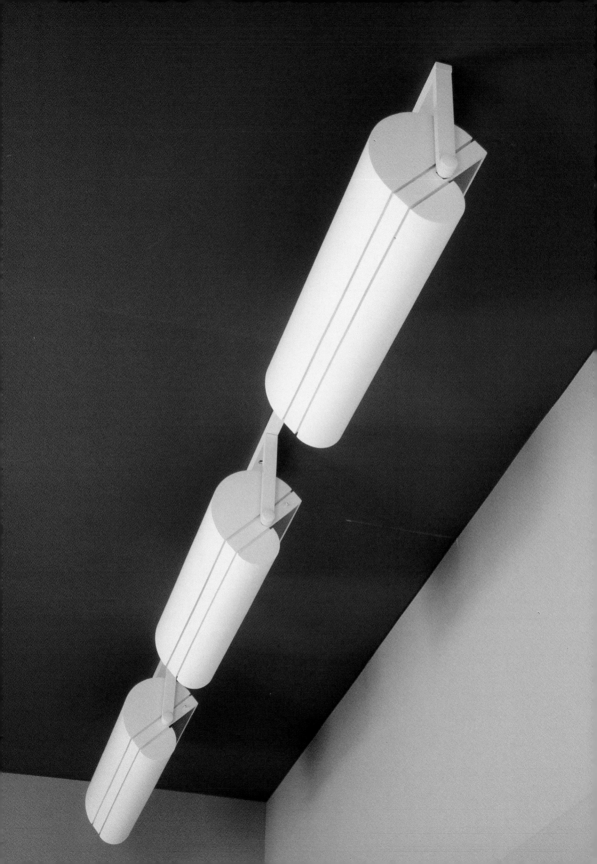

TRIAD WALL-ILLUMINATION FIXTURE 1983

The ceiling-mounted fixture is designed to wash a wall with light and to allow adjustment of the light pattern and placement; thus each of the fixture's three elements can be individually rotated. The diameter of the smaller half-cylinder of each of the elements is the radius of the larger. The incandescent tube is centered in the small half-cylinder, which is lined with reflective material. The light is then re-reflected from the larger half-cylinder, softening and distributing the light and masking the glare of the incandescent tube.

The piece was designed to be produced in enameled steel, but fabrication difficulties mandated a change to sheet brass. The mounting frame is a rectangular steel tube. The entire assembly is coated in a matte enamel.

Jacobson admits that the half-cylindrical form may not be perfectly suited for precisely even light distribution; the ideal might be something closer to a parabola. But he intended from the outset that the fixture should have a bold and memorable sculptural presence, and the pure Euclidian geometry is beautifully suited to that purpose.

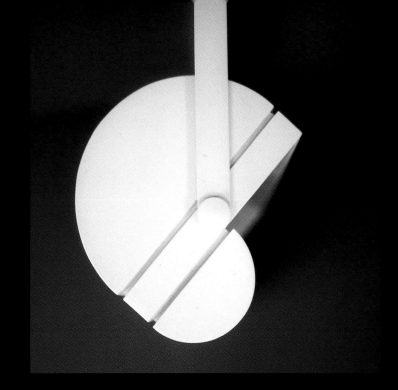

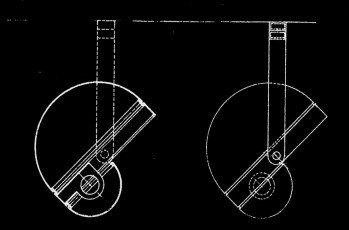

Design Evolution

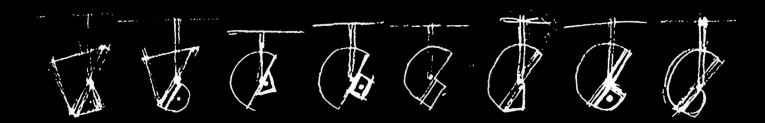

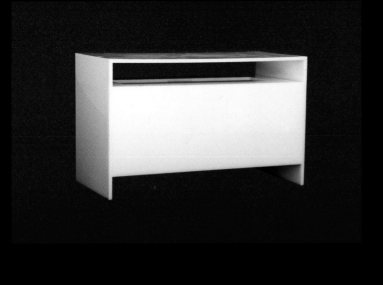

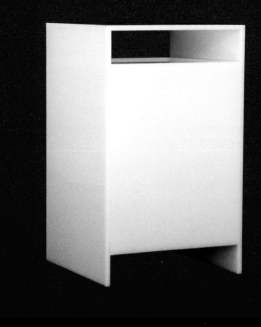

THE WELDED ACRYLIC SERIES:
MODULAR TABLE LAMPS 1975

This series is designed around the characteristics of white acrylic sheet material. The material can be cut to almost any two-dimensional shape and can be formed to singly and doubly curved surfaces, but it is most straightforwardly and economically used in planar form. The acrylic is unaffected by time or the heat of the lamp. Neighboring pieces are joined by "welding": a solvent bonds the mating surfaces to make the two pieces one; the strength of the joint is that of the material itself.

These three welded acrylic lamps are dimensionally coordinated for use singly or in groups; the base dimensions are 7-$^1/_2$" by 7-$^1/_2$" and 7-$^1/_2$" by 13-$^1/_2$", with varying heights. Their crisp rectilinearity derives from Jacobson's acceptance of the material's planarity and its method of assembly. The open areas at top and bottom provide essential air circulation to ventilate the lamp's heat.

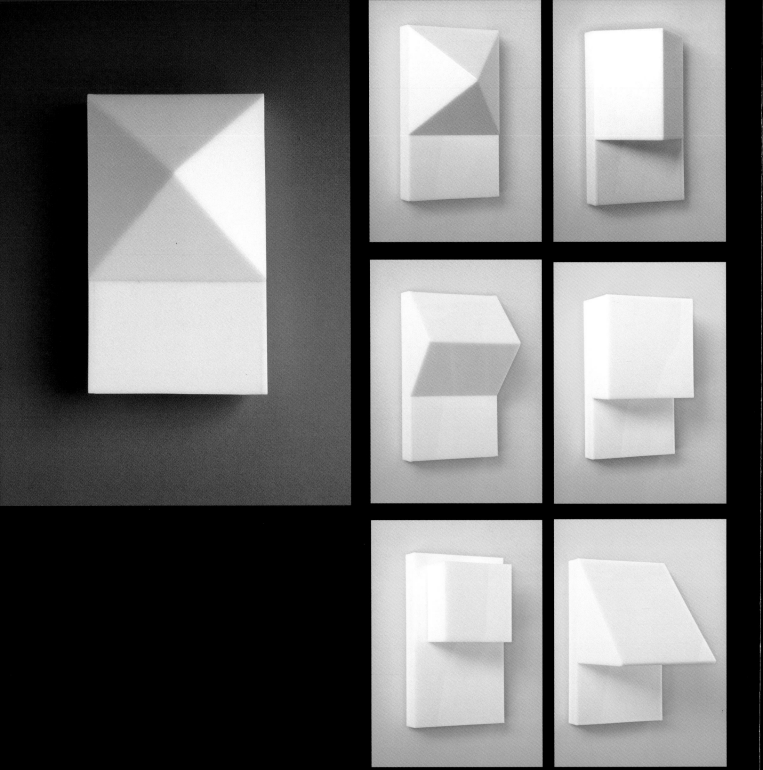

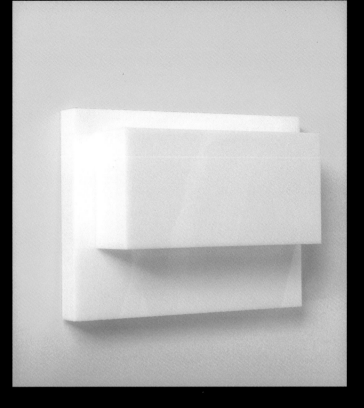

THE WELDED ACRYLIC SERIES:
WALL SCONCES 2003

The sconces are designed to be used singly or in group-ings of like or different units. The overall outline of each of the six variations is a Golden Section rectangle, 7 by 11-$^3/_8$ inches. From that rectangular prism, a form pro-trudes based, in all cases, on a perfect square. Jacobson is fond of citing a quotation from Albert Einstein: "Make all things as simple as possible, but no simpler." The pre-cept applies to much of Jacobson's design work and is es-pecially evident in this sconce series. These sconces are, in a sense, conceptual cousins to the wall triad: in each case the geometric image is both elemental and memorable.

Representative Groupings

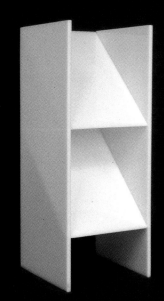

THE WELDED ACRYLIC SERIES: ORTHOGONAL AND TRIANGULAR READING LAMPS 2004

The upper sloping plane of this lamp group masks the light source; the lower reflects light outward. The single and double orthogonal units are modularly coordinated and can be used in multiples. The triangular unit provides a wider spread of light and can be tucked into an interior corner, or it can serve as an end piece or an external corner for orthogonal groupings.

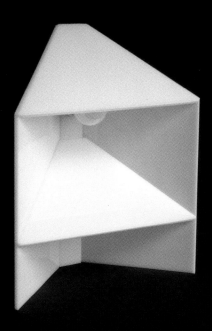

THE WELDED ACRYLIC SERIES:
THE STAR PENDANT 2003

In preparing for their marriage, Phil and Effie had acquired a blown opal glass pendant dining lamp of roughly teardrop shape, of which, over the years, they became very fond. When a migrating crack destroyed it, Jacobson began this pendant design as a replacement and a reprise. His self-imposed challenge was to recall a continuous doubly curved form by means of a design employing only purely planar elements

The design comprises seven cascading tiers of planar facets; horizontal acrylic diaphragms connect the tiers. The facets, however, present a formal problem that does not arise when a fixture is circular in plan. If the lamp is to hang, as is usual, within a rectilinear space and is simply a square in plan, the facets must either align with the wall planes or lie at 45 degrees to them; any other relationship will be seen to be askew. But given conventional methods of suspending such a fixture, its orientation cannot be controlled; therefore a square-plan geometry can never be satisfactory. Jacobson found, however, that the plan figure of an eight-pointed star presents a sufficient number of facets that the eye does not identify any of them with the context; therefore a fixture with that plan geometry can satisfactorily hang in any space, in any orientation.

The diminution of the upper tier is determined by a 45-degree angle; the lower tier by a 30-degree angle, so the profile approximates the teardrop form of its glass predecessor. The configuration yields intriguing planar and volumetric effects that change with movement of the viewer: seen from a point somewhat lower than the lamp's center the geometry of the its lower tier can seem identical to the upper, yet when seen from above the lower tier can disappear entirely. The top and bottom of the lamp are open for air circulation and bulb replacement.

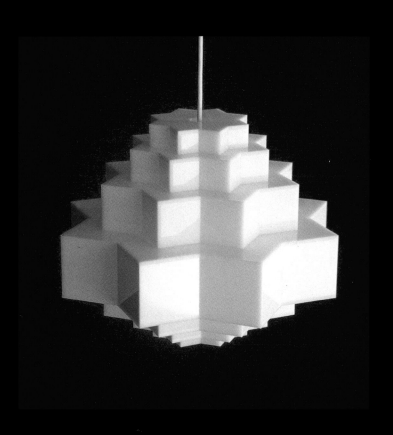

Jacobson's interest in jewelry had its beginnings in 1952-53, before, during, and after his graduate work at Liverpool, when he visited crafts schools in England, Germany, Denmark, and Sweden. Numerous subsequent visits to Mexico brought to his attention the exuberance of its folk arts and especially the silverwork of central Mexico: "Here I first encountered the wide spectrum of artifacts that are possible employing this marvelous material."

His introduction in 1968 to Hannes Niemi, of the Helsinki College of Crafts and Design, marked his first encounter with the actual making of jewelry and, specifically, the crafting of silver, with which Niemi had then been involved for more than forty years. Jacobson has since worked with many other craftspersons, from whom he has learned much; their experiences have informed his designs.

He has said that of all his cognate design interests, the design of jewelry is farthest removed from the considerations that play a central role in furniture design, lighting fixtures, and architecture; Jacobson describes jewelry as an almost wholly artistic venture. As sources for his inspirations he cites geometric principles, analogies to forms in nature or industry, recollections of past architectural examples, and, occasionally, his unprompted imagination. Nevertheless, his jewelry is unmistakably the work of an architect. It has, in every case, a governing geometry as

rigorous as that imposed by any architectural system of columns and beams; its lines are the lines of architectural materials. It is, quite obviously, architecture as jewelry or jewelry as architecture. In many examples there is too, paradoxically, something of the early modernist discipline of economy of materials, a clear intention to accomplish an ornamental purpose with a minimum of formal means.

And there is a deeper parallel between jewelry and architecture, because jewelry must also, in a more modest way, respond to the laws of physics. Gold is one of the heaviest of the elements, silver is not far behind, precious stones are dense concentrations of mass, and diamonds are created by the compacting force of unimaginable pressures. Yet jewelry is supported by insubstantial cloth or the human neck, or wrist, or finger—so in all cases the weight of the piece must be considered in creating the design. It must especially be considered when the piece is to be relatively large in overall dimension or is to include some relatively massive element. The distribution of the masses—the physical balance of the piece—must also be thought through. The center of gravity of a brooch must lie quite near the plane of the garment it adorns to avoid pulling the garment forward; a bracelet or ring must not be so asymmetrically weighted that it tends to rotate on the wrist or finger. The mass of a brooch or pendant must be approximately symmetrical about the

SOURCES OF INSPIRATION

left to right:

Clay Pots, Oaxaca, Mexico, 1971;
Floor Mosaics, Venice, 1988;
Eroded Beach Rock, Matia Island,
Washington, 1973; Sibelius
Monument, Helsinki, 1968; Islamic
Fabric, 1982

point of suspension, if the piece is to hang as intended. A
pendant's breadth must be sufficient to ensure that it lies
flat against the garment, resisting any tendency to twist.

Two very different production techniques pervade the design process from the outset.

Fabrication entails the assembly of standard manufac-
tured bar, tube, and sheet materials to create the finished
piece, a process that allows some detail changes during
the process. Fabrication tends to encourage somewhat
simpler forms, since the complexities obtained in casting
by simply pouring the liquid into the mold must be built
up by assembly, in a fabricated design, piece by tiny piece.

left to right:

Giant's Causeway, Northern Ireland,
2003; Teotihuacan Pyramid Complex
Mexico, 1971; Ears of Corn, Pike
Place Market, Seattle, 1986; Church
of San Biagio, Montepulciano, Italy,
1993; Mitla Ruins , Mexico, 1971

Casting for Jacobson's work is typically done by the lost-
wax process, in which a wax simulation of the actual finished piece is created, the form is cast around the wax
simulation, and the wax original is melted away. Liquid
silver or gold is then poured into the void; when solidified
and withdrawn, it then will have the shape intended for the
piece. Casting enables the development of complex forms,
but with limited ability to modify those forms after casting.

And, of course, the two techniques can be combined.

left to right:

Building Façade, Prague, 1998;
Motorcycle Cylinder Head; Oak Bark
San Juan Island, Washington, 1991;
Wood Pile, Finland, 1968; Wall Cap,
Oaxaca, Mexico, 1971

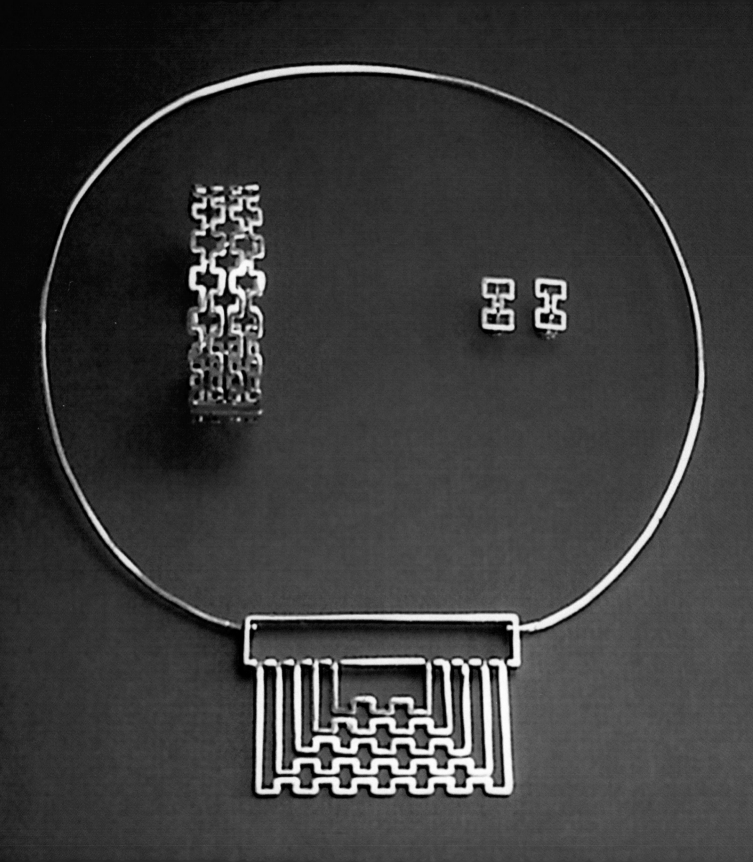

FABRICATED DESIGNS:
THE SERRATION SET 1968

This set marks Jacobson's first venture into the design of jewelry. He developed these pieces during his sabbatical in Helsinki, in concert with Finnish jeweler Hannes Niemi, from whom he learned the elements of fabrication processes and the characteristics of the jeweler's materials. The pieces of this set are of a silver alloy (not sterling) specially developed by Niemi for the project.

The design of a set such as this presents a challenge that is not present in the case of a single piece, since the basic conceptual idea here must be capable of successful adaptation to pieces of several sizes and configurations.

The serrated motif of these pieces was inspired by minerology and biology, as well as by Islamic geometric patterns to which Jacobson was introduced in his travels in Africa and the Near East. It was also inspired by certain examples of Italian architecture; Jacobson cites the seemingly serrated entablature of the Church of San Biagio at Montepulciano and the typical geometric themes of architect Carlo Scarpa. The serrations create a complex family of tiny facets which catch the light in continually changing relationships; the piece sparkles and glitters not only of itself but in response to the human movement inherent in its use. In that sense this set introduces a characteristic that will permeate all of Jacobson's jewelry designs to follow.

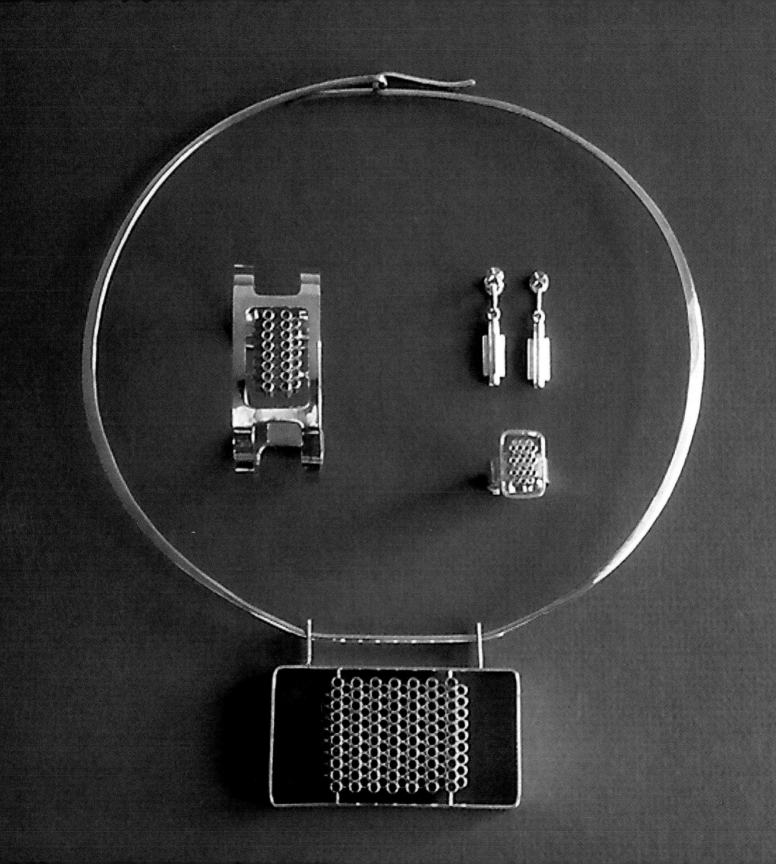

FABRICATED DESIGNS:
CLUSTERED CYLINDERS 1970

These four pieces derive their theme from the Sibelius memorial in Helsinki, honoring the greatest of Finnish composers; the memorial is a vertically disposed but seemingly random assemblage of cylinders suggesting organ pipes at monumental scale. But Jacobson's design in this case is much removed from its inspiration. The highly structured pattern of cylinders arranged in parallel rows is given complexity and enrichment by the projection of alternating rows at alternate dimensions. Unlike the following suspended-cylinders set, these cylinders are fixed and unmoving. Yet the pattern they create is also dynamic, since, as in the case of the serrated set, the relationships between the cylinder clusters seem to shift and change as the complex geometry catches the light differently with each change of viewing angle.

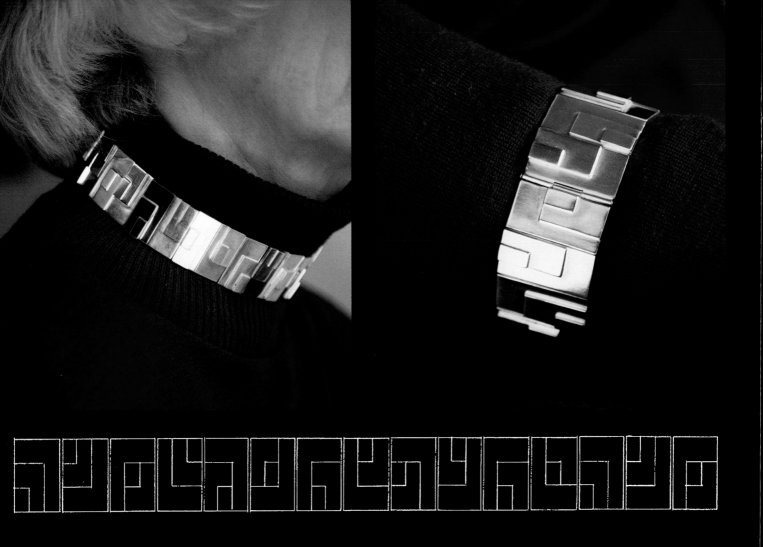

FABRICATED DESIGNS:
COLLAR AND BRACELET 1973

System of Geometric Organization

A rectilinear proportion is repeated at two-thirds and one-third of the linear dimension, then is layered in silver sheet stock as a three-dimensional record of two-dimensional pattern. The great variety of arrangements creates the unexpected richness of the design.

FABRICATED DESIGNS:
THE PENDANT SET 1980s

Variations on a theme: three pendants, from yellow-gold bar stock, with birthstones of tourmaline and topaz, for Jacobson's daughter, his daughter-in-law, and Effie.

FABRICATED DESIGNS:
SUSPENDED CYLIINDERS 1969

The inspiration in this case was Japanese wind chimes. The cylinders are of solid stock and are suspended to move individually within the frame; the material is solid sterling silver.

FABRICATED DESIGNS: FIN PENDANTS 2005

These pieces were inspired by mechanistic industrial images—automobile grilles, fin-tube heating systems, the finned cylinders of air-cooled engines. The fins alone carry the design in the single-fin iteration; in the double-fin version the paired central fins grasp the gemstone of lapis lazuli. In each case, the fins project forward in the clustering zone for sculptural depth. The tapering edges of the projection soften the effect and add an element of sculptural complexity; they also prolong the dominant light-catching surfaces, thereby intensifying the dynamic brilliance of the piece.

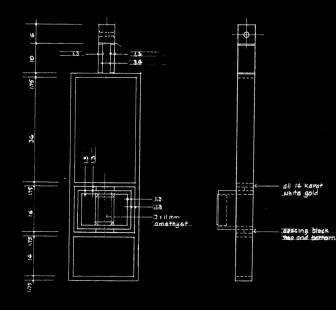

FABRICATED DESIGNS:
THE AMETHYST PENDANT 2005

Nested rectangles of white gold frame a dark amethyst. This piece illustrates the ubiquitous problem of the chain attachment, which invariably demands considerable study. In this case, does the design suggest one pierced vertical bar? Two? A vertical prism? Is the attaching element to be echoed by a similar feature at the bottom of the piece, or not? As often happens in design, the final choice, in this case a pierced cube poised on paired vertical bars, seems right and proper, but was far from obvious at the outset.

HYBRID DESIGNS: GOLDEN SECTION
PENDANTS AND TORC 2004

These pieces also commemorate a wedding anniversary, the Jacobsons' fiftieth. The five dark-blue sapphires in each piece, one for every decade, are complemented by the cast white-gold frames. The frames are proportioned as Golden Section rectangles, and the fabricated fins that hold the gems are half or full Golden Sections. The 45-degree corner gussets express the cast nature of the frames and add a complexity to the otherwise pure geometry.

These pieces, and the one to follow, are among many in which Jacobson has used the Golden Section as a geometric discipline. The Golden Section is a rectangle so proportioned that side A is to side B as side B is to the sum of A plus B; the resultant proportion is 1:1.616. The proportion interests Jacobson, as an architect, in part because of its long and distinguished role in the history of Western architecture. The most familiar example is found

CAST DESIGNS:
PRIMARY FORMS PENDANTS 1994

The three primary forms are common to the symbolism
of many early cultures. Here they are developed in depth
as the three primary volumes: the cube, the sphere, and
the tetrahedron. The white gold is a neutral companion to
the stones of red coral, blue lapis lazuli, and green chryso-
phase. Jacobson's intention was to use stones in the primary colors, for obvious reasons, but no yellow example
seemed suitable.

CAST DESIGNS:
THE HEXAGONAL RING 1980

Geometric games, circles, and hexagons: a pure cylinder
poises a hexagonal prism, both in white gold; within the
prism a circular diamond recalls the cylinder.

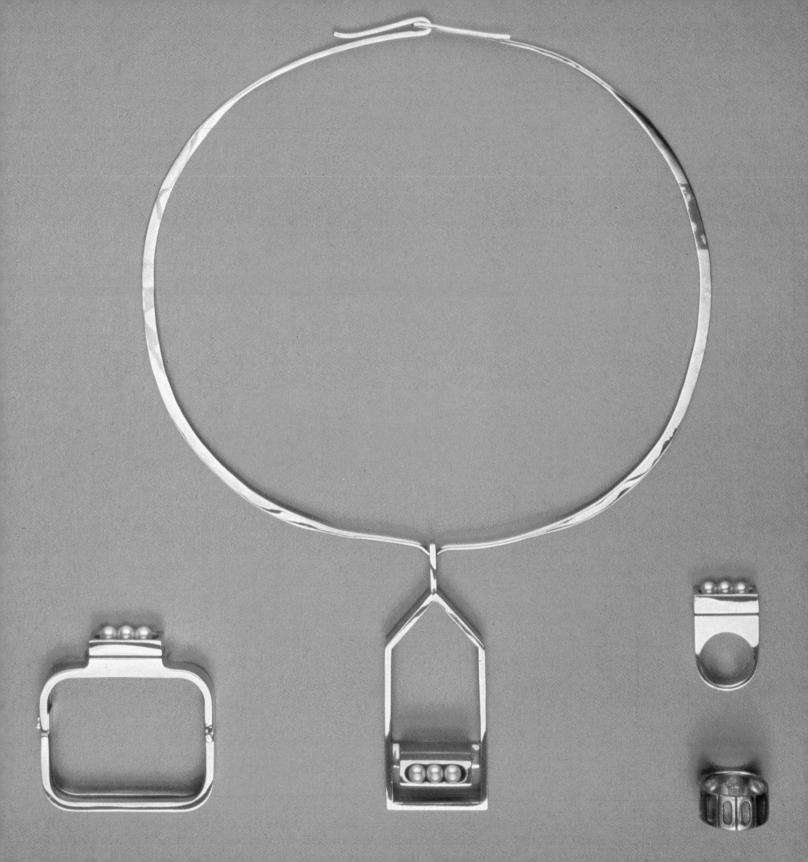

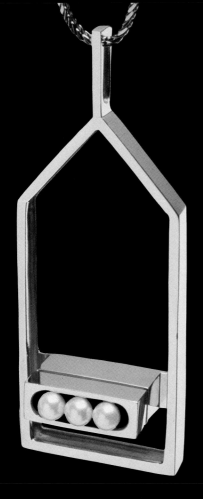

CAST DESIGNS: THE PEARL CLUSTER SET
1979

This series derives from Jacobson's long-standing interest in natural examples of clustered repetitive elements—flower stamens, ears of corn, peas in their pods. The white pearls are set in yellow-gold castings.

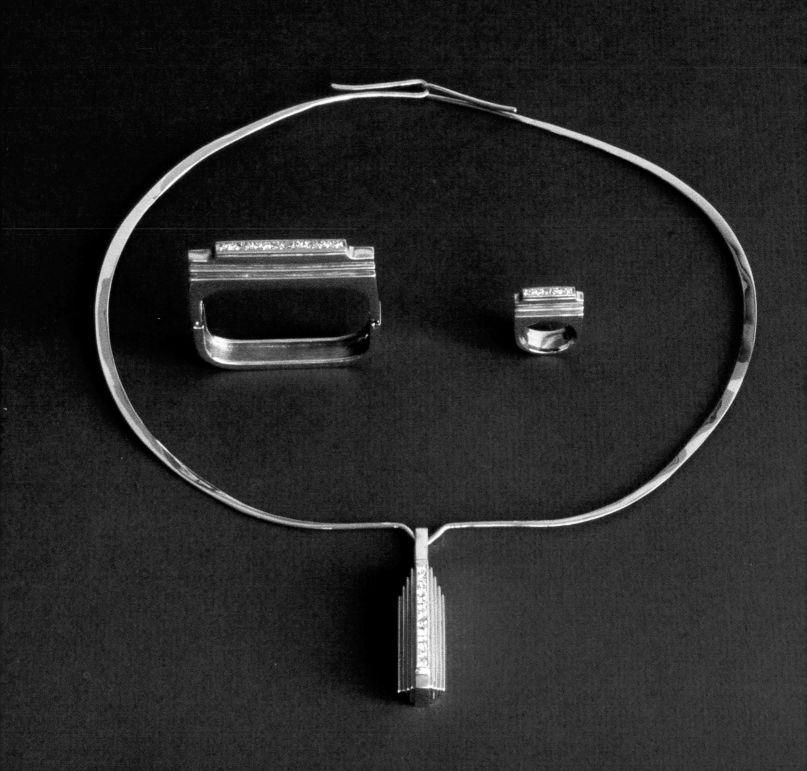

CAST DESIGNS:
THE DIAMOND STRIP SET 1989

The pendant, bracelet, and ring were created to celebrate the Jacobsons' thirty-fifth wedding anniversary. Square-cut diamonds, thirty-five in all, are set in rows within the parallel steps of the platinum prisms; the complementary geometry of diamonds and setting achieves a remarkable elegance. Architecture is again the inspiration and reference, in this case the tradition of monumental stairs. Jacobson had especially in mind the sequence, by the Danish architect Jørn Utzon, that comprises the majestic podium of the Sydney Opera House.

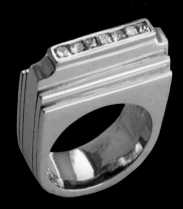

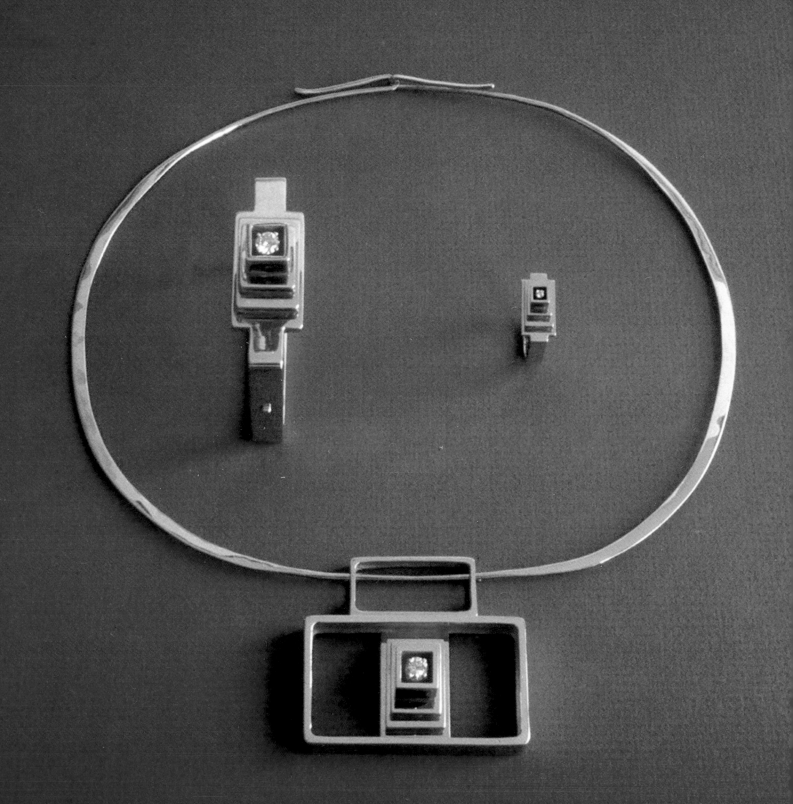

CAST DESIGNS: THE ZIGGURAT SET 1984

One of Jacobson's most dramatic architectural memories is of a low-altitude flight over the pyramids of Teotihuacan; another, equally vivid, is of his first view of the Egyptian pyramid at Saqqara, the world's first example of architecture in stone. Those pyramids are stepped, however, and so are formally more analogous to the stepped forms of the early Mesopotamian ziggurats—man-made monumental brick mounds. Hence Jacobson's naming of this set. The massive presence of these pieces, in yellow gold with single crowning diamonds, is homage to those architectural memories.

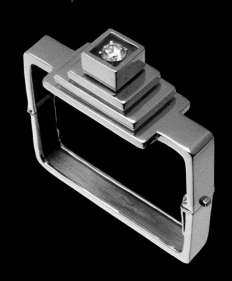

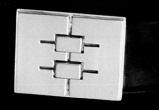

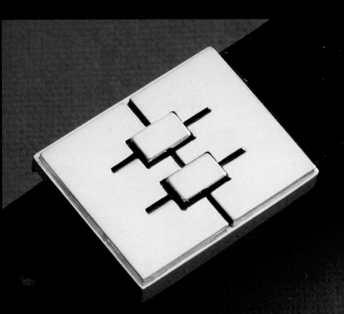

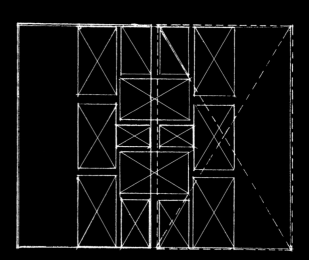

Interrelationship of Golden Sections

an "E." That was the starting point. Jacobson mirror-imaged the two letters, then let them move about, asymmetrically at first, then symmetrically. Simple became more complex, then simpler again, but always with allegiance to the clarity of the opposed "E"s. With continuing iterations, an emerging pattern began to suggest the proportions of the Golden Section, in which, as we have seen, Jacobson has an ongoing interest. More precise studies revealed that the entire design could resolve into implied and explicit Golden Section rectangles.

And so the piece came together—one of the most personal and charming of Jacobson's efforts, an intriguing melding of the unique and the universal, the sophisticated and the simplistic, the obvious, the subtle, and the obscure.

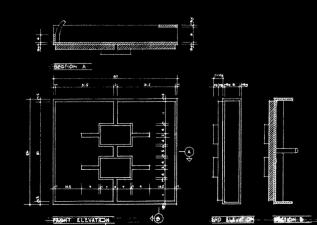

SECTION A

FRONT ELEVATION

END ELEVATION

SECTION B

V SERVING PIECES

Jacobson has designed a number of other small items for both personal and institutional use; these two serving pieces are among his favorites. Their design inevitably involves a consideration of their balance in the hand; they must also suggest to the user the way they are to be held and offer a hand grip that is both secure and unstressful. The tea and coffee set also mandates a formal theme capable of application to the wide variety of sizes and

Design Alternatives

A COCKTAIL AND BUFFET TRAY 2004

**Designed in collaboration with Alex Osenar, the tray is
made of injection-molded polycarbon. It will hold a wine
glass, a tumbler, or a beverage can, and an array of hors
d'oeuvres, its sophistication impervious to the bustle of a
stand-up cocktail party.**

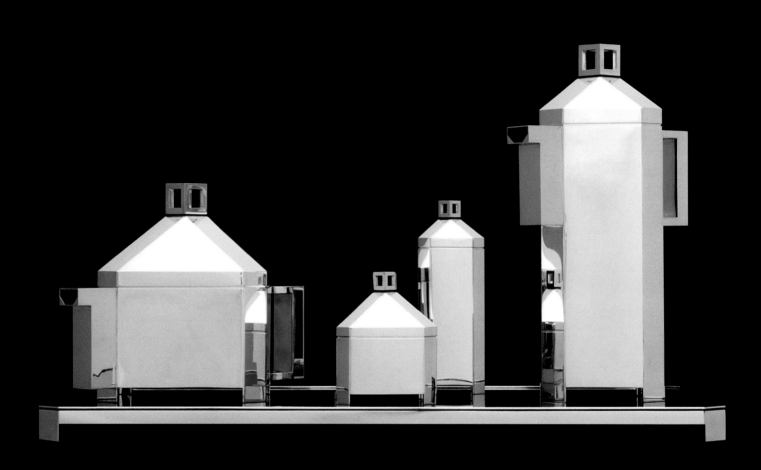

THE TEA AND COFFEE SET 1986

The set comprises tea and coffee pots and cream and sugar containers, to be carried on the black Delrin plastic floor of the silver-framed tray.

The lower volumes of the serving pieces are irregular octagonal prisms, symmetrical about four axes. All handles and spouts project from the narrow vertical faces as a secondary family of rectangular prisms. The crown of each piece is generated by folding the four broad sides of the octagonal prism inward as triangular panels and folding the four narrower sides inward as rectangles that intersect as a square, from which a pierced cube rises to become the knob. The tray echoes, more simply, the formal themes of the pieces it is designed to carry.

The pieces are fabricated from solid sterling silver plates soldered and polished; all seams are "chased" as infinitesimal chiseled valleys, to enrich the planar intersections while resolving any imperfections of alignment.

The architectural character of these pieces is self-evident; so too is the strangely abstract power of their simple yet complex geometry. Seen apart from context, they can seem an almost surreal metaphor for some scaleless futuristic city or, equally, some yet-undiscovered ancient monument field.

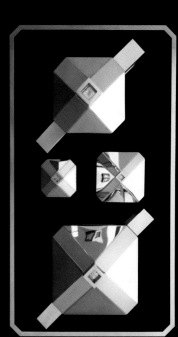

VI AFTERWORD

Phil Jacobson has long since retired from both practice and teaching, but he has not retired from the realm of design this book portrays. Dividing his time now, between Southern California and the Puget Sound region, he continues to design lighting fixtures, especially desk lamps, many of which his son Erik "tests" and is notably reluctant to return. And Jacobson continues to design jewelry— more of it than ever, in fact—for friends, for family, and above all for Effie, who continues to find delight in each new addition.

Grant Hildebrand taught at the University of Washington from 1964 through 2000, as Phil's colleague and friend. In that time, among other things, he produced three books, and now, retired, he spends most of his time producing more of them. For his wife Miriam, he recently asked Jacobson to create an anniversary pendant, which she now wears with much pleasure and pride.

Cathedral Roof
Budapest, 1998

BIOGRAPHICAL DATA

PROFESSIONAL CAREER

University of Washington Professor of Architecture and Urban Design and Planning	1962–2000
TRA	1959–1992
Architecture Engineering Planning Interiors, Seattle	1955–1957
John Carl Warnecke Architect, San Francisco	1957–1959
John W. Maloney Architect, Seattle	1953–1955

PROFESSIONAL AND CIVIC INVOLVEMENT

Northwest Institute for Architecture and Urban Studies in Italy Board of Trustees	1990–1994
AIA Seattle Senior Council President	1985
Seattle Architectural Foundation Board of Trustees	1985–1991
AIA Seattle Board of Directors	1970–1972
Washington State Council of Architects President	1965–1966
Northwest Trek Foundation Board of Trustees	1987–1994
Pilchuck Glass School Board of Trustees	1982–2001
Northwest Seaport Board of Directors	1980–1982
City of Seattle Landmarks Preservation Board	1976–1980

Store Entrance Door Surround
Rome, 1986

EDUCATION

Finnish Institute of Technology, Helsinki M. Arch (Licensiata)	1969
University of Liverpool, Urban Design and Planning	1952–1953
Washington State University B. Arch. E. (Honors)	1952

AWARDS

AIA Seattle Chapter Medal	1994
Finnish Society of Architects Silver Award	1992
Government of Finland Knight First Class, Order of the White Rose	1985
American Institute of Architects Fellow	1973
Fulbright Senior Research Fellow, Finland	1968–1969
Fulbright-Hayes Scholar, England	1952–1953
Washington State University Outstanding Alumnus, Sigma Tau	1964

Member of six university honor societies

CREDITS

Many talented craftspersons have been involved in the execution of Phillip Jacobson's designs.

FURNITURE

Bill Walenta	The Modular Seating System
Rudy Mayer Karl Mayer Joe Cortez	The Dining Table, Sideboard, and Coffee Tables
Curtis Erpelding	The Laminated Wood Series: Coffee Table Chair 1 Chair 2 Counter Stool

LIGHTING

Tom Wimmer	Institutional Fixtures
Gary Rubins	Triad Wall Illumination Fixture
Eric Ganung	Modular Table Lamps The Star Pendant
Steve Sissenstein Carlos Garnica	Wall Sconces and Reading Lamps

Oak Leaves
San Juan Island, Washington, 1971

JEWELRY

Hannes Niemi	The Serration Set The Suspended Cylinders
Margie Ogle	The Clustered Cylinders Set The Collar and Bracelet
Lorri Ferguson	The Pendant Set The Fin Pendants The Amethyst Pendant The Diamond Strip Set The Primary Form Pendants The Golden Section Pendants Effie's Belt Buckle
Daniel Louis	The Pearl Cluster Set The Ziggurat Set The Hexagonal Ring

SERVING PIECES

Todd Feldsted Barbara Alder	The Tea and Coffee Set

PHOTOGRAPHS

All photographs are by Phillip Jacobson, with the exception of the following:

Alan Abramowitz	p. 4
James Housel	p. 2
Charles Pearson	p. 5
Christian Staub	p. 5
Hugh Stratford	p. 5
Tom Upper	p. 4
Photographer Unknown	pp. 2, 49, 50, 61, 62

ACKNOWLEDGMENTS

The exhibition from which this book evolved and the book itself were originally suggested by Elaine Latourelle; she, Jim Nicholls, and Vikram Prakash encouraged and assisted the project from its inception. Michael Duckworth of the University of Washington Press sympathetically guided the book through its many phases of development and production. And Rolf, Chris, Erik, and Effie have been at Phillip's side, and in his thoughts, through all the many decades in which he has created the good works celebrated herein.

Publication of *Elegant Explorations* was supported by John and Dorothy Anderson, T. William Booth Architect, Arne Bystrom FAIA Architect, Dale Chihuly, Curtis Erpelding, C. David Hughbanks, Wendell Lovett FAIA Architect, Allen and Elizabeth Moses, Nathaniel and Faye Page, and an anonymous donor.

Bill Booth graciously offered his assistance in garnering support for this publication. His good efforts were essential in realizing the venture.

We thank everyone involved.

G.H. and P.J.